Farm Animals

Cows

Rachael Bell

Heinemann Library
Chicago, Illinois

Designed by AMR
Originated by Ambassador Litho
Printed in Hong Kong/China.

04 03 02 01 00
10 9 8 7 6 5 4 3 2 1

Library of Congress Cataloging-in-Publication Data
Bell, Rachael.
    Cows / Rachael Bell.
        p. cm. – (Farm animals)
    Includes bibliographical references and index.
    Summary: Describes the habits and behavior of cows and how they
are kept and cared for on a dairy farm.
    ISBN 1-57572-529-0
    1. Dairy cattle Juvenile literature. 2. Cows Juvenile literature.
[1. Cows. 2. Dairy farming.] I. Title. II. Series: Bell, Rachel.
Farm Animals
SF208.B37 2000
636.2'142—dc21   99-42837
                                                             CIP

Acknowledgments
The Publishers would like to thank the following for permission to reproduce photographs:
Agripicture/Peter Dean, pp. 11, 15, 18, 19; J. Allan Cash, pp. 25, 27; Heather Angel, pp. 8, 26; Anthony
Blake Photo Library, p. 22; Anthony Blake Photo Library/Maximillian, p. 23; Chris Honeywell, p. 29;
Hutchison Library, p. 24; Images of Nature/FLPA/R. Bird, p. 4; Images of Nature/FLPA/L. Lee Rue, p. 7;
Images of Nature/FLPA/Peter Dean, p. 10; Images of Nature/FLPA/Daphne Kinzler, pp. 12, 13,14;
Images of Nature/FLPA/G. T. Andrewartha, p. 20; Images of Nature/FLPA/F. de Hooyer, p. 28; Lynn M.
Stone, pp. 6, 16; Tony Stone Images/Gary Vesta, p. 5; Tony Stone Images/Dante Burn-Forin, p. 9; Tony
Stone Images/Andy Sacks, p. 21.

Cover photograph reproduced with permission of Robert Harding Picture Library.

Our thanks to the American Farm Bureau Federation for their comments in the preparation
of this book.

Every effort has been made to contact copyright holders of any material reproduced in this
book. Any omissions will be rectified in subsequent printings if notice is given to the Publisher.

Some words are shown in bold, **like this.** You can find out what they mean by looking in the glossary.

Contents

Cow Relatives

Cows are important farm animals. They are **raised** all over the world. There are many different kinds of **cattle.** Many farmers keep black and white cows for their milk.

Long-horned cattle were brought to the **Americas** by Christopher Columbus. The **descendants** of these animals went into Mexico and Texas. Today some farmers raise Texas longhorns.

Welcome to the Farm

This **dairy** farm has about one hundred cows. Most of these are Holstein cows. The cows are black and white. The farmer milks them twice every day.

Half the land is **pasture**, for the cows to eat. The farmer also grows corn, wheat, and **barley.** He turns the corn into **silage** for the cows to eat in winter. He sells the wheat and barley.

Meet the Cow and the Bull

tail   **udder**   jaw

Female **cattle** are called cows. Cows usually have one calf each year. A cow cannot give milk unless it has had a calf.

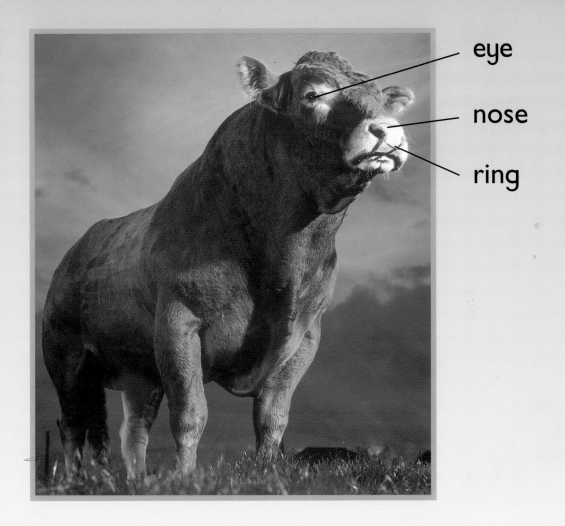

eye

nose

ring

Male cattle are called bulls. Bulls are larger than cows. Sometimes a bull has a ring in its nose so the farmer can lead it. A bull can weigh as much as a small truck.

Meet the Babies

Baby **cattle** are called calves. A calf stays with its mother for the first three or four days to get the best of her milk. The calves then drink milk from a machine.

When a calf is a few hours old, the farmer puts a tag on each ear. The numbers on the tag tell which calf it is and which farm the calf is from.

Where Do Cows Live?

From spring to autumn, cows stay outside in the **pasture**. When the cows have **grazed** the grass, the farmer moves them into a new field.

In the winter, most cows are kept inside a barn to protect them from the cold. Every day the farmer puts fresh straw in the barnyard. He scrapes it out every three weeks.

What Do Cows Eat?

In summer, cows eat fresh grass. They pull grass into their mouth with their tongue. They chew and swallow it. Later they **chew the cud**.

In the winter, the farmer feeds the cows with grass or corn **silage** and a **mineral lick**.

Staying Healthy

Cows like to be together in a **herd**. They know to follow each other in **order** for milking and feeding. The farmer always keeps the cows in this order.

Cows often lick themselves or each other to rub off insects. They **flick** flies off with their tails. They also like to scratch on a fence-post or a tree.

How Do Cows Sleep?

Cows lie down on their side to sleep. If they are are outside, they choose a **sheltered** place and lie with their backs to the wind.

Each day, cows spend about eight hours eating, eight hours **chewing the cud,** and eight hours resting.

Raising Cows

Some farmers milk their cows by hand. Most farms use milking machines to milk groups of cows at one time.

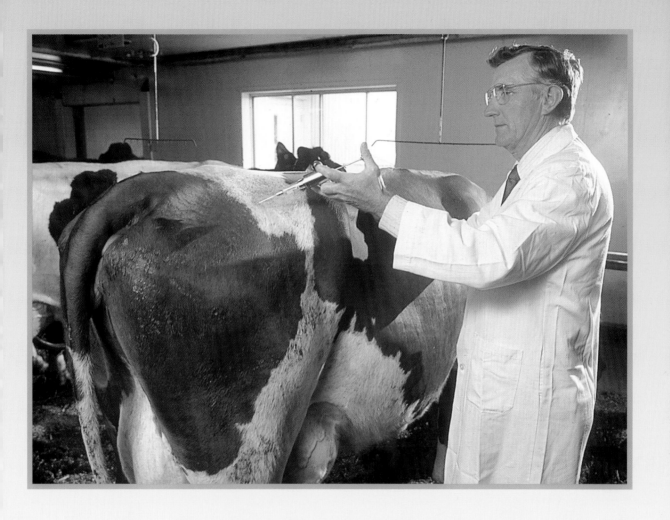

The farmer and his family clean the barnyard and **milking parlor.** In winter, they also feed and **bed** the cows in the covered yard. If a cow gets sick, the **veterinarian** may come to help.

How Are Cows Used?

The cows on a **dairy** farm are kept for their milk. In one year, a cow can give more than twenty thousand glasses of milk! The milk is put in jugs or made into yogurt, cheese, butter, and ice cream.

Some farms keep **cattle** for their meat. The meat from cattle is called beef. It is sold in different cuts, such as steak, ground beef, or roasts.

Other Cattle Farms

There are some very large **cattle** ranches in Australia and the **Americas**. **Herds** of cattle are loose on miles of open land. **Stockmen** and cowboys round them up on horseback.

In Australia, cattle **stations** may be very far from the nearest city. The cattle are taken to market in huge **road trains** or special trucks.

Cattle at Work

In some countries, **cattle** are used to pull heavy loads and farm equipment. In **Asia,** cows pull **plows** through flooded fields so the farmer can plant rice.

Some cattle are kept for **rodeos**.
Cowboys ride horses and use a **lasso**
to catch calves. This does not hurt the
calves, though.

Fact File

 A calf can walk and drink milk from its mother almost as soon as it is born. A cow knows her calf by its smell. Her sense of smell is better than her eyesight. Because of this, you often see a cow nose to nose with her calf.

 A **dairy** cow eats grass every day. It can eat its own weight in grass every week!

 There are almost two billion beef and dairy **cattle** in the world.

Some leather comes from cattle. The skin of a cow is called a hide. Leather from hides is used to

make clothes, shoes, purses, and gloves.

Cattle have 32 teeth. They have eight teeth in the front part of their lower jaw. There are twelve teeth in the back of their upper jaw. There are twelve more teeth in their lower jaw.

Glossary

Americas North, Central, and South America

Asia largest land area of the world that includes countries such as China, Vietnam, and Thailand

barley plant that can be made into cereal

bed put fresh straw down on the floor for the cattle to lie on

cattle baby and adult cows and bulls

chew the cud to bring food back up into the mouth from the stomach and chew it again

dairy cows kept for their milk

descendant an animal that came from certain ancestors

flick to knock something away with a quick movement of the tail

graze to nibble or eat grass or plants

herd name for a group of cattle

lasso rope used to catch cattle

milking parlor building where cows are milked

mineral lick block of substances such as salt that animals lick to help them grow

order way in which animals follow one another

pasture fields of grass for animals to eat

plow large blades pulled through soil to turn it over

raise to feed and take care of young animals or children

road train many containers hooked together, pulled by a truck

rodeo shows where cowboys show how they ride and rope

shelter protected from bad weather

silage winter food for animals made by cutting plants and storing them

station very large ranch in Australia

stockmen people who look after animals, which are also called stock

udder part of an animal's body where milk is stored

veterinarian doctor for animals

More Books to Read

Aliki. *Milk from Cow to Carton.* New York: HarperCollins Children's Books, 1992.

Brady, Peter. *Cows.* Danbury, Conn.: Children's Press, 1996.

Hansen, Ann L. *Cattle.* Minneapolis: ABDO Publishing Company, 1998.

McDonald, Mary A. *Cows.* Chanhassen, Minn.: The Child's World, Inc., 1997.

Index

School Dist. 64
164 S. Prospect Ave.
Park Ridge, IL 60068